Trash Attack!

Garbage,

and

What

We Can

Do

About It

By Candace Savage
Illustrated by Steve Beinicke

earth care books

Douglas & McIntyre
Toronto/Vancouver

Douglas & McIntyre Ltd.
585 Bloor Street West
Toronto, Ontario M6G 1K5

Canadian Cataloguing in Publication Data

Savage, Candace, 1949-
  Trash attack!

(Earthcare books)
ISBN 0-88894-826-3

1. Refuse and refuse disposal — Juvenile
literature.
I. Beinicke, Steven, 1956- . II. Title.
III. Series.

TD792.S38 1990   j363.72'88   C90-094476-5

With special thanks to Vanessa Alexander.

Design by Michael Solomon
Printed and bound in Hong Kong

Contents

The Problem

You are about to read something disgusting.

In just one year, most people throw away about 80 cans full of garbage. Eighty big stinking garbage cans, cram-jammed, packed to the brim.

We're not talking about 80 cans for each family. It's 80 for every mother, 80 more for each dad, 80 cans for each and every kid.

Eighty garbage cans, placed side by side, would probably cover the floor of your living-room. That's just one person's share for a single year. Think of it: a room full of gum wrappers, juice tins, soggy paper towels, broken bottles, dirty diapers, wrecked toys, moldy sandwiches—paper, plastic, glass, metal and food—all squished together in a rotting, mixed-up mess.

And here is something even worse. Almost everything buried in that slimy heap is still perfectly good. Toys that could be fixed; paper, metal, plastic and glass that could be recycled. Even most of the food scraps could be reused, if they were composted. With a little care, they could make your garden grow! When we throw things in the garbage, all this goodness is wasted.

The things we put in the garbage originally came from the Earth. Every time you throw away a piece of paper, for example, you are throwing away a tiny bit of forest. Maybe it was a twig where a bird liked to sing; or a leafy branch that made shade for a wood violet.

The garbage you throw out in a single year probably weighs more than SEVEN times as much as you do.

To make paper, people cut down trees. Sometimes whole mountainsides are cleared at the same time. We don't like to think about what happens when we do this, to the deer and rabbits, the frogs and spiders. We don't like to think about how long it will take for the trees to grow back. We don't even want to think of them as trees anymore, after they're chopped down. We call them logs and take them to a paper mill.

At the mill, the wood is mashed into a thick gray soup called pulp. At some mills, chemicals like laundry bleach are added to the pulp to make the paper white. When the pulp is pressed dry, it becomes paper. The water that drains out of the paper is flushed away into our lakes and rivers, or is allowed to drain into the ground. This water is filled with these chemicals, and some of them are poisons. The air that comes from the mill's smokestacks also contains dangerous chemicals.

If we keep on
At this rate
Things aren't
*   going*
To be that great!

For every piece of paper that is made this way, a bit of chemical-filled water has been added to the Earth, and a whiff of chemical-filled smoke has been added to the air. So when you throw away a piece of paper, you are also throwing out a cup of clean water and a breath of fresh air.

TRASH TIMES-TABLE

To find out how many garbage cans you may have filled in your life, take your age and multiply by 80; or look up the answer on this chart. You can also use the chart to do a Trash Tally for your parents or friends. How much garbage will you likely create if you live for 80 years?

AGE	NUMBER OF CANS
1	80
5	400
6	480
7	560
8	640
9	720
10	800
11	880
12	960
20	1,600
30	2,400
40	3,200
80	6,400

If you put 6,400 garbage cans side by side in a row, they would stretch out for more than 3 kilometers! That's about 2 miles or 20 city blocks.

Become a Trash Attacker!

Join the **Trash Attack!** and earn a special certificate! Here's all you have to do. Watch for this picture throughout the book: . It marks **Trash Attack!** activities. Photocopy the **Trash Attack!** certificate on the opposite page. Each time you complete a **Trash Attack!** activity, color one of the 's on your certificate. When you have finished six of the activities, you will have earned your **Trash Attack!** certificate.

This is to certify that

_____

*Is a Qualified Trash Attacker
and is doing her/his part
to keep the Earth
green and growing.*

To qualify as a Trash Attacker, complete six of the Trash Attack activities in this book. Look for this picture:

Each time you finish one of the activities, color one of the 🗑's on your certificate.

Trash Attack!

The Great Garbage Guesstimate

Eighty cans a year is the average amount of garbage one person creates, which means that some people throw out less, some even more. If you live in North America, you probably throw out more than someone who lives in Europe. Europeans, in turn, produce more garbage than Africans.

It's easy to figure out your own garbage score. Just count the number of cans or bags of garbage that your family throws out each week. Then do a little figuring. If math gives you a headache, ask someone for help!

Name: _____ Date: _____

In one week, my family threw away ____ cans or bags of garbage. There are ____ people in my house. If we each threw out about the same amount, then my share of the garbage is the number of cans or bags divided by the number of people. That comes to ____ cans for me each week.

There are 52 weeks in a year: 52 times my weekly number is ____ cans or bags. That's how much garbage I make in just one year.

AFRICA   EUROPE   NORTH AMERICA

Once a forest has been cut, it will never be quite the same again. But at least new trees can be planted. Because they can be replaced, we say that trees are "renewable."

Many things we take from the Earth *cannot* be replaced—metals, minerals and oil. They are "non-renewable," which means when we use them up, they are gone. Millions of years ago, the Earth was very different than it is today. Deep oceans covered places that are now dry land. Tiny animals swam in these waters, and when they died, their bodies sank to the bottom of the seas. Silt and sand settled on them, layer upon layer, until the tiny squashed bodies slowly turned into oil. All the oil on Earth was made this way.

We use oil for almost everything. We burn it as gasoline in our trucks and cars. It heats our homes, schools, offices and factories; it powers machines and generates electricity. It is made into plastic, clothing and medicines.

The garbage we throw out today may be in our great-great-great-great-great-great-great-great-great-grandkids' way!

The Garbage Glossary

RESOURCE WORDS

Greenhouse effect: A warming of the atmosphere around the whole Earth. This is caused by gases in the exhaust from cars, furnaces, factories and other industry—anything that burns oil, gasoline, natural gas or coal. The gases form a layer above the Earth that holds the heat in, just like the glass in a greenhouse.

Non-renewable resources: Things we use that we cannot replace, such as oil (petroleum), aluminum and iron. When they're gone, they're gone.

Renewable resources: Things that we can use without using them up. Plants grow back; animals have babies; the wind always blows.

We are using oil so fast that some people say it will be gone in fifty years. Others say thirty. And as we burn it, we are putting huge amounts of exhaust gases into the air, causing the Greenhouse effect. This is altering the atmosphere of the whole Earth and causing the climate to change.

When you toss out a plastic bottle, you are throwing away the oil in the plastic and the fuel that was burned to make and transport the bottle. You are throwing away the clean air that was dirtied by the plastics factory and the transport trucks.

Every time something is manufactured, it uses up energy and other resources, and pollutes the water and air. By buying things, throwing them out and buying again, we are hurting the Earth. It is time for a change!

Tracking Down Your Trash

Garbage even causes us problems when we try to get rid of it! No matter what we do, it keeps coming back to us, in the water, in the soil, in the air.

Did you ever wonder what would happen if no one came to take your trash away, or anybody else's? What if people simply threw their garbage out the window and left it to rot in the streets? A few hundred years ago, that's exactly what happened in cities across Europe and North America. Fat garbage-fed rats scurried underfoot, and people died of terrible sicknesses. That's why garbage collection was invented.

Garbage collection solved many problems but it also caused new ones. The biggest one is this: we don't have to think about our garbage anymore. The garbage collectors toss it into their trucks, rattle off down the street, and —poof!— the whole mess just seems to disappear.

Truck it far
Or truck it near
Garbage does not
Disappear!

The fact is, of course, that the garbage inside the truck goes right on being garbage. The same old problem, in a new place. Where do you suppose it all goes? When you throw something away, where is "away"?

Trash Attack!

Calling All Detectives!

Find out where your garbage goes. Ask your parents or teachers. Phone your local government offices. Talk to the garbage collectors. Ask questions; snoop around. It's your garbage, after all. You have a right to know!

LANDFILLS

Most of our garbage is either buried or burned. Places where it is buried are called dumps, or landfills. Dumping garbage is quick, easy and not too expensive; but it takes up a lot of space. In North America, where most garbage is taken to dumps, towns and cities are running out of room for their trash. In Los Angeles, in New York, in Toronto, in two years or ten, the dumps will be full. People call this the Garbage Crisis.

Most of the land near cities and towns is already filled up with houses, farms, parks and roads. There are no places left for garbage dumps. Some cities have tried to solve this problem by opening landfills out in the country, if the rural people will agree. (Would you agree, if the dump was to be near your house?) Others try to ship their garbage to poor nations, hoping that someone will sell them disposal space. But these days nobody wants to take in other people's trash, not even if they are going to be paid.

When do people think YOUR dump will be full? How old will you be then? Where will your garbage go afterwards?

Garbage dumps can be dangerous. When rain falls on a landfill, water seeps into the piles of

trash. If the water trickles over dirty diapers or soiled tissues, for example, it picks up germs. If it flows through spilled paint, motor oil or household cleaners, it picks up dangerous chemicals. All this guck—called "leachate"—flows out the bottom of the dump into the ground beneath.

It doesn't stop there. The polluted water soaks into the land, finding places were it can trickle through soil or porous rocks. On and on it flows, through cracks and around boulders, sometimes trickling into fast-flowing underground streams. This is the water we draw from wells and that flows into our rivers and lakes. Far away from the dump, fish swim in it; people drink it. Once the groundwater has been contaminated, it is almost impossible to clean, because there is no way to get at it.

A garbage dump can pollute the groundwater for hundreds of years. This is because everything we throw into a dump eventually breaks down, or decomposes.

Some things break down fast. They rot, which means they are digested by tiny creatures such as bacteria and molds, and turned into food for plants. Food scraps, grass clippings, leaves—

I don't even Want to think it Tell me I don't Have to drink it!

things that were once alive—all break down this way; so does paper, which is made from trees. We say these things are "biodegradable."

Plastics, glass and metal cans are "non-bio-degradable," and they break down much more slowly. Nobody knows how long it will take for plastic to break down, because plastic has only been in use for about fifty years. Some people think that it may take 500 years for the plastic in a disposal diaper to disappear completely. If dangerous chemicals were put into the plastic when it was made, they may trickle into the water for a long, long time.

Adventures of a Garbage Barge

By 1987, Islip, a city in northeastern New York State, was running out of room to dump its garbage. So one spring day, a mountain of baled trash weighing 2,700 tonnes (3,000 tons), or more than 350 large elephants, was loaded onto an ocean-going barge. Pulled by a tugboat, it headed out of the harbor to look for a landfill where it could leave its load.

Day after day, the tug chugged southward along the east coast of the United States, stopping at towns and cities along the way. But nobody would accept the garbage. The tug sailed on to Mexico and across to Belize and the Bahamas. But no one there would let it unload either. Eventually, after a journey of 164 days, the tug returned home, still pulling its large and smelly load.

Cities in northeastern New York now send most of their garbage by truck to landfills in other parts of the United States. Some loads travel 1,400 kilometers (870 miles) before they're dumped! People pay about $300 a year each just to have their garbage trucked.

Another problem with dumps is that they catch on fire very easily. When things rot, a gas called "methane" is produced. This is the same as the natural gas that many people use to heat their homes. If methane collects in an enclosed space—in the middle of the dump, surrounded by piles of trash, for example—it may explode. If this happens, it lights the garbage on fire, and polluted smoke from the burning trash billows into the air.

Well-run dumps have vent pipes to carry the gas away so it cannot explode. Some cities do even better: they collect the methane, clean it up a bit, and sell it as fuel. This lets them save a non-renewable resource (natural gas that has been pumped out of the ground) by replacing it with a renewable resource (garbage gas).

The Latest in Landfills

New! Improved! The very best garbage dumps now wear plastic pants! Before a new landfill is opened, the whole area is covered with plastic or a special kind of clay. This seals the ground so that water from the garbage cannot seep away. Pipes are installed to collect the polluted water so it can be cleaned, and ditches are dug around the landfill to catch spills, in case there are accidents. New landfills are now being built this way, which is good news. (Liners cannot be put in landfills that are already in use.)

But there is bad news as well. The liner in a landfill soon gets buried under mountains of trash. What if, after a while, it develops a leak or a crack? If people wanted to fix it, they'd have to move all that garbage first! Landfill liners are a new idea, and nobody knows how well they are

**Garbage Dumps
Get Low Grades**

D—*for air pollution, because dumps can catch on fire and give off dangerous smoke.*

D—*for water pollution, because chemicals from the garbage soak into the earth.*

D—*for usefulness, because dumps eventually become full and have to be closed.*

D—*for waste, because things just lie there in the dump until they are ruined.*

going to work. Will they last until you're grown-up or until you have grandchildren? Nobody can say.

There's only one kind of garbage dump that is totally safe—an empty one. The less we throw in the garbage, the better things will be for all of us!

The Garbage Glossary

DUMP WORDS

Biodegradable (Buy-oh-de-GRADE-a-bull): Something that rots.

Groundwater: Water flowing under the ground.

Landfill: A garbage dump, or tip.

Leachate (LEECH-eight): The polluted water that seeps from a garbage dump.

Methane: Natural gas that is produced when things rot.

Non-biodegradable: Something that does not rot.

Trash Attack!

How to Make Gorgeous Green Glorp

Here's an experiment that will help you discover what rots and what does not. All you need are a few bits of garbage and a place where you can leave them for a while.

- First, choose one or two scraps of food and paper—an apple core, a crust of bread, a piece of paper.
- Choose one or two items made of plastic, metal or glass—a plastic bag, metal lid, a glass bottle.
- Lay all the items outside on moist dirt, in a spot where no one

is likely to bother them. Cover them to keep off dogs and cats. If you can't put them outdoors, put them on a baking sheet, moisten them with water, cover them and tuck them out of the way.

- Check them after a week, after two weeks, after a month. Which ones change? What happens to them? What do you think would happen after a year, or a century?
- When things rot, they break down into nutrients that plants need to grow. Nature gets rid of garbage by using it over again.

For more fabulously "rotten" ideas, see Compost on page 50.

Does Your Garbage Go to a Low-Down Dump?

A "good" landfill is like a "good" nightmare. It doesn't exist. Still, some are better than others. Here is a list of four things that can be done to make a dump safer. How many of them are done at your landfill?

■ The garbage is covered with soil each day to keep away animals and to prevent trash and bad smells from blowing around.

■ Hazardous Wastes, such as batteries and paints, are kept separate from other garbage. Things that can be recycled, such as pop cans, glass bottles and paper, are removed before the garbage is dumped.

■ Methane is piped out of the garbage so that it will not catch on fire. The gas is refined and sold as fuel.

■ The landfill is lined with plastic or clay so that polluted water is less likely to seep into the ground.

If your landfill gets a low rating, don't worry. Do something! Talk to your friends and teachers. Tell your parents. See if your coaches and club leaders will help. *Find someone else who cares.* It might be one person; it might be your whole class. Together, you could

■ gather more information
■ publish a newsletter
■ design posters
■ write letters to the mayor
■ write dumpy poems and trashy songs.

INCINERATORS

If your trash does not go to a landfill, it is probably burned in an incinerator. From the outside, an incinerator looks like a factory with large smokestacks; inside it has a big pit, into which garbage trucks dump their loads of trash, and a huge furnace in which the garbage is burned.

Incinerators are popular because they save space: a mountain of garbage turns into a hill of ash. In Europe and Japan, where people ran out of room for landfills long ago, most garbage is burned. Many new incinerators are also being built in North America.

In an incinerator, your 80 cans of garbage burn down to 8 cans of ash, a plume of smoke and a blast of heat. In some incinerators, water is piped through the walls of the furnace, where the heat of the fire turns it to steam. The steam is then piped away and used to warm buildings or make electricity. In this way, your 80 cans of garbage produces enough energy to light a lamp, night and day, for four months. Incinerators that do this are called Energy-from-Waste Facilities, or Resource Recovery Facilities.

But burning garbage to produce energy is not as good as it may sound. In fact, some people find it downright frightening. Here's one reason why:

Most of the things we throw in the garbage have hurt the Earth in some way. To grow our food, we have polluted the soil with pesticides. To make and transport manufactured goods such as paper and plastic, we have polluted the water and air and used up petroleum, which we cannot afford to waste. We are making the Earth a worse place to live. For our own sake, and the sake of all living

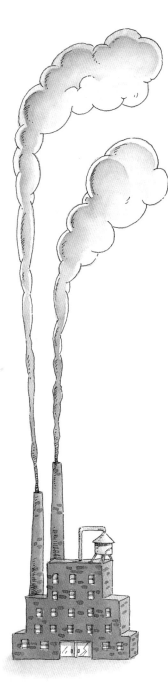

things, we have to stop using and wasting so much. We must make less garbage.

But Energy-from-Waste Facilities need garbage—piles and piles of it—day after day. They are hungry mouths always waiting to be fed. If we depend on garbage for heat or electricity, it becomes harder for us to cut back on the amount we throw away. If a town or city has spent all its money to build incinerators (which are very expensive), there may be nothing left to spend on recycling. Energy-from-Waste Facilities may keep us from meeting the challenge, which is to keep our planet alive!

Incinerators cause other problems as well. Burning garbage produces two kinds of ash: "bottom ash" (which settles to the bottom of the furnace) and "fly ash" (which flies up the smokestack). If Hazardous Wastes go into the incinerator—ordinary things like paints, batteries and

More energy can be saved by recycling than by burning garbage.

antifreeze—both kinds of ash are dangerous. Even materials that were safe when we used them may become harmful when they are burned. Printers' ink, paper, plastics and metal cans all contain poisonous chemicals that may be freed when they are burned. New dangers may also be created when chemicals combine.

Most of this nasty stuff stays inside the incinerator. Some of it falls to the bottom. The rest is caught by special "scrubbers" and filters in the smokestack. Their job is to trap fly ash and clean up the fumes before they leave the chimney.

Making an incinerator burn safely is difficult, because the garbage is not always the same. In the autumn, for example, people may throw away truckloads of leaves. Because they are damp, the leaves lower the temperature of the fire in the incinerator. Dangerous chemicals, which are destroyed inside the incinerator when the fire is hot, escape from the smokestack when the fire is cooler. When that happens, the incinerator must be shut down immediately.

Even the best-run incinerators can cause problems. Some fumes and ash always escape, adding to big-city smog and to acid rain. Even worse, they release whiffs of poisonous metals (mostly lead) and other pollutants into the air. These fall on nearby fields and gardens and get into our food. There's not enough to make most people sick right away, but what happens if you breathe it and eat it for ten or twenty years? What will it do to trees, fish, birds and other living things?

The ash that is caught inside the incinerator is a worry, too. What should we do with it? Nowadays, most of it just goes to dumps, many of which are unlined. What is to keep the poisons in the ash

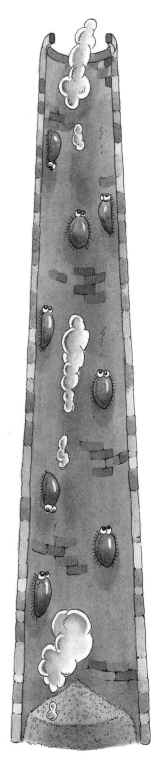

Incinerators Get Bad Marks

from leaching into the groundwater? In some places, people are more careful. They believe that the ash is unsafe and treat it like an industrial Hazardous Waste. In parts of Germany, for example, fly ash is packed in drums and stored in underground caves, along with poisonous chemicals from factories. In Japan, fly ash is mixed with cement, formed into bricks and buried in lined landfills.

To learn about Hazardous Wastes, read "Poison Trash" on page 26.

D—*for air pollution, because they release ash and fumes through their smokestacks.*

D—*for water pollution, because they produce huge blocks of dangerous ash that are often buried in landfills, where the poisons in the ash can leach into the groundwater.*

D—*for expense, because they are very costly to build.*

D—*for waste, because they need mountains of garbage to burn. They do not help us learn to throw less away.*

Garbage University

In Germany people have to study for two-and-a-half years before they are allowed to work at a garbage incinerator. In Switzerland, incinerator managers and supervisors also must take special courses. In most other countries, they are not required to do so.

Go to Jail! Go Directly to Jail!

Every country has rules about how much pollution an incinerator may put into the air. In Germany, an incinerator manager who knowingly lets his plant break the air-pollution rules is automatically sent to jail for two years.

How "Hot" Is Your Incinerator?

If your garbage goes to an incinerator, you'll want to know that it's the best possible. Here's a list of four things that can be done to improve an incinerator. How many are done where you live?

■ Paper, aluminum, glass and other materials that can be recycled are removed before the garbage is burned. Hazardous Wastes are taken out, too. This leaves only a small amount of garbage, which can be burned to produce heat or energy.

■ The "scrubbers" in the chimney, which filter the smoke, are new; and the smoke that leaves the chimney is analyzed constantly. Information about the gases and chemicals in the smoke is displayed outside the incinerator where people can see it as they go by.

■ The staff at the incinerator have all had special training to work there. When they can tell that the incinerator is not operating safely, they shut it down.

■ The blocks of leftover ash are buried in a lined industrial landfill or some other secure place.

If you don't like
What you find out
Tell your friends
And start to SHOUT!

The Garbage Glossary
INCINERATOR WORDS

Acid rain: Rain that has been changed by exhaust from cars, houses and industry. Acid rain kills trees and fish.

Bottom ash: The ash that falls to the bottom when garbage is burned.

Dioxins and furans (die-OX-ins, FYR-ans): Families of chemicals, some of which are very poisonous. Dioxins and furans are often found in the smoke from incinerators.

Energy-from-Waste Facility: An incinerator that uses the heat from burning garbage to make steam. The steam is used to warm buildings or make electricity.

Fly ash: The ash that goes up the smokestack when garbage is burned.

Hazardous Wastes: Garbage that contains dangerous chemicals.

Incinerator (in-SIN-er-ay-ter): A special furnace for burning garbage.

Resource Recovery Facility: Same as Energy-from-Waste Facility.

Scrubbers: Devices in the chimney of an incinerator that help clean up the smoke.

Smog: Polluted fog that hangs over cities because of exhaust from cars, houses and industry.

POISON TRASH!

Hazardous Wastes contain chemicals that make people sick. If they get into the water or air, they harm plants and animals as well. Whether your garbage is buried in a landfill or burned in an incinerator, Hazardous Wastes mean trouble. Some of this dangerous garbage comes from our homes.

Before you throw something in the garbage, think: "What if this stuff finds its way back to me through the water or the air? Would I like to drink or breathe THIS?"

Here is a list of household products that should NEVER go into the garbage or down the drain.

Paint and paint thinners
Oven cleaners
Batteries
Floor and furniture polishes
Used motor oil and other fluids
   from cars and trucks
Swimming pool chemicals
Photographic chemicals
Silver polish and other
   metal polishes
Toilet and drain cleaners

Bleach
Scouring powders
Herbicides (to kill weeds)
Pesticides (to kill insects, rodents
   and other pests)
Disinfectants (to kill germs)

Trash Attack!

Hazards at Home

Every year, most families throw out enough Hazardous Waste to fill a kitchen sink! Use the list above and check for hazards in your home. What will you do with the leftovers and dirty containers when you're finished with them?

Some communities have regular depots where people can take their Hazardous Wastes for safe disposal. Others have special days when Hazardous Wastes are picked up. Some have no plan for dealing with Hazardous Wastes at all.

How are Hazardous Wastes handled in your community? Ask your family and friends, or call city hall. If you do have access to a depot or a special pick-up, be sure to use it!

Trash Attack!

No Depot for Hazardous Wastes?

If your community does not have a Hazardous Waste depot or pick-up, why not ask for one? It's easy. All you need are several friends, enough pens to go around and a writing pad. Write letters to your local government and your newspaper. Tell them that Hazardous Wastes are a danger to people and other living things. Tell them that more and more communities have plans for storing or disposing of them safely. Yours could be the next!

If writing letters doesn't appeal to you, why not make posters about Hazardous Wastes and display them at a meeting of your local council? Or you could make puppets and let them talk to the council for you. Have fun!

Trash Attack!

Clean Cleaners

Ever cleaned your bathtub with a lemon? It's fun, it works and—best of all—there are no Hazardous Wastes! The next time it's your turn to clean the bathroom, why not try some of these safe ways to make it shine?

- For windows and the mirror: Instead of glass cleaner, dissolve a little baking soda in water. Wash, rinse and polish as usual.

- For the sink and tub: Take half a lemon, dip it in borax and rub it on stains. Then rinse with clean water. (You can buy borax in the laundry section of your supermarket.)
- For the counter: Try a little vinegar mixed in warm water. Wash and rinse.
- For the toilet: Use baking soda or a little mild dishwashing detergent. Scrub the toilet with the brush.

- For the drains: Pour 25 mL (2 tbsp) baking soda down the drain. Then pour in 25 mL (2 tbsp) vinegar. Leave for 15 minutes, then run hot water down the drain.

For more suggestions, talk to your grandparents or older friends. When they grew up, people didn't use so many strong chemicals around the house. See what they remember. Sometimes old ideas are the newest thing around!

The Solutions

The best way to
Clean up this
mess
Is to start
wasting
Less and less.

"Bigger!" "Faster!" and "More!" are not words that we can live by. Our throw-away habits are changing the Earth too much, too quickly. The good news is that we don't have to do it anymore.

There are three words that will help us to live in a better way. They are called the "3 R's": Reduce! Reuse! and Recycle!

REDUCE!

Reduce means to buy less and throw less away. The best way to cut down on the amount we waste is to stop buying things we don't need in the first place. This means that we have to pay attention when we go to the store.

Let's say you have just received your allowance, and you are dying to get inside a store. You want to shop, shop, shop. You want to spend, spend, spend. Whoa! Slow down. As you make your selection, give yourself a moment to think.

- Do I need this thing, or even want it? Will it really make me happy? Will I still want it a year from now?
- How long am I going to use it?
- Is it likely to break right away? Can it be repaired? Will it be good enough to give to someone else when I'm through with it?
- Do I have something at home I could use instead? Could I borrow one or buy one second-hand?

- What is it made of? Did it cause pollution when it was made, or will it cause problems if it's buried or burned? Is it biodegradable? Can the materials be recycled or reused?
- How is it packaged? Does it have more containers than it needs? What am I going to do with the wrappers?

A cookie wrapped in foil
Sitting on a plastic tray
Is not the kind of packaging
That really makes my day!

Packages—the bags, boxes, bubbles, buckets, jars and tins that things come in—are a big source of waste. If you sorted the garbage in your trash can, about half of it would be packaging. In a year, you may fill 40 garbage cans with packaging alone. Some packaging is important: you can't take milk home in your hand. But many things in our stores have more wrappers than they need. Toy manufacturers, for example, know that children like colored labels, fancy boxes and shiny cellophane wrap. They hope that lots of glitzy packaging will make you buy their product, even if you don't want or need whatever is inside. To make things even worse, they charge you for the extra packaging, as part of the price of the toy. As soon as you get home, all those glamorous wrappers go straight into the trash.

The Solutions / **31**

Some kinds of packaging are worse than wasteful; they are dangerous. High above the Earth there is a special layer of air called the "ozone layer," which protects us from bad sunburns and skin cancer. In the last few decades, people have begun to manufacture and use chemicals that float to the top of the atmosphere and destroy this natural sunscreen. These chemicals are called CFCs, and they have been widely used in two kinds of packaging—aerosol spray cans and plastic foam containers.

Aerosol spray cans are the kind that you don't have to pump. You just push on the button and whatever is inside—whipped cream, air freshener, furniture polish—spurts out. But something else comes out as well: chemicals that were put in with the product so that it would spray properly. Until recently, almost all aerosols used CFCs for this purpose, and some still do today. To avoid CFCs, look for cans that are marked "Flammable" or "No Ozone Damage." These labels are a sign that other chemicals have been used instead of CFCs. Better yet, avoid aerosols altogether. Even many non-CFC aerosols contain substances that can seriously harm your health. When you push the aerosol button, the contents come out in a fine mist, which you breathe in. Anything in the aerosol goes straight into your lungs and bloodstream. Not only that, but the cans cannot be recycled or reused, and all aerosols, full or empty, are explosive. This means they can blow up in an incinerator or garbage truck, or even when they are buried in a landfill. So look for products that come in refillable bottles, with pumps that you work by hand.

Plastic foam is the material used to make hamburger boxes, egg cartons, meat trays and disposable coffee cups. When some of these foams are made, CFCs are used to puff up the plastic. Other chemicals, which are thought to be safer, are now often used instead. Unfortunately, there is no easy way to tell a "safe" foam from one that was made with CFCs. So ask questions and look for labels and signs. If you aren't sure that a foam is safe, don't use it. Look for products that are packaged in cardboard instead.

*Do not spray
Those aerosols
Or we will all
Need parasols!*

Trash Attack!

Packaging Patrol

If you find something in the store that is over-packaged, write to the company that made it. (The address should be on the package; otherwise ask the store clerk for help.) In your letter, say that you did not buy the product, and explain why. Companies that make things want us to buy them. If we don't and tell them why, they will pay attention. Writing letters can make a difference.

Trash Attack!

Container Countdown

How many containers does your garbage container contain? (Try saying that ten times, fast!) While you're waiting for your tongue to untangle, here's an easy way to answer the question. Instead of putting all your trash in one bag, set out two. Put the packaging in one: jars, tins, plastic tubs, wrappers, all that sort of thing. Put everything else in the other. Do this for one week. How many bags did you fill with packaging? How many would you fill in one year? (Multiply the number for one week by 52.)

RULES FOR REDUCING

DO buy large-sized packages. (One big package makes less garbage than two small ones. It also usually costs less.)

DO shop for quality. Buy things that are made to last.

DO rent or share things you seldom use.

DO buy things "loose" instead of in packages. Buy pens and pencils from a bin, instead of in bubble packs. Shop at bulk food stores, where you can take your own containers and fill them with things like sugar, raisins and rice.

DON'T buy things you don't really want or need.

DON'T buy things that can't be fixed.

DON'T buy things that were made to throw away, such as disposable flashlights and cameras, and pens that won't take refills.

DON'T eat in restaurants that use disposable dishes.

DON'T buy anything that is over-packaged.

DON'T use aerosol sprays.

DON'T use plastic foams unless you are sure they are free of CFCs.

The Garbage Glossary

SAVE-THE-EARTH WORDS

CFCs: Chemicals used in some aerosol spray cans and plastic foams that eat away at the ozone layer. CFC stands for chloro-fluorocarbons (KLOR-o-FLOR-o-kar-buns).

Ozone layer: A special layer of air, high in the atmosphere, that protects us from sunburn and skin cancer.

Recycling: Saving used materials and sending them off to be processed into useful goods.

Reducing: Buying less and throwing less away.

Reusing: Fixing and altering things we already have so that we can continue to use them.

Trash Attack!

Tips for Trash Attackers

For each of these suggestions you follow, give yourself ONE credit toward your **Track Attack!** certificate.

■ If your school cafeteria uses throw-away dishes, take your own plate and cutlery—or see if you can persuade your school to use real dishes.

■ At home, use cloth napkins instead of paper ones, and rags instead of paper towels.

■ If you have a baby in your house, talk to your parents about using a diaper service instead of disposable diapers. It's cheaper, just as convenient, and much less wasteful! People in the United States alone throw out 18 billion disposable diapers every year)

■ Take your own shopping bags to the grocery store. In France, people do it all the time; anyone who forgets is charged $1.50 for a plastic bag.

■ Check out the garbage at your favorite fast-food place. If you love their food but not their trash, write and tell them so. Find a new favorite place—one that makes less garbage. Or follow the example of a Canadian class that went to McDonald's and took their own cutlery, cups and plates!

Trash Attack!

Do You Take Garbage for Lunch?

Here's an easy way to check your lunch box for unnecessary waste.

■ As you eat your lunch, pile the garbage neatly on the table. Look at it carefully. Which of these items are in your garbage heap— lunch bag, paper napkin, plastic bag, twist tie, drinking straw, drink box, raisin box, plastic wrap from a cookie or treat? What else is there?

■ Count the number of things you are throwing away. Multiply the total by the number of students in your class or in your school. That will give you an idea of the number of items that go into the garbage at lunch time each day.

■ In a year, there are about 200 school days; so multiply by 200 to get the number of trash items you throw out in one year. Is it a little bit, or a lot?

■ Now look at the things in your garbage again. How could you have brought the same lunch, without all the trash? For example, maybe you could use a lunch box instead of a paper bag, or carry your drink in a Thermos instead of a box. You could cut your own cheese instead of buying wrapped slices, or buy raisins in bulk and bring a handful in a little jar. Or perhaps you could pack your sandwich in a reusable container instead of a plastic bag.

■ Write down your ideas, take them home and show them to your parents. Ask them to help you make your lunch trash-free!

Trash Attack!

Think Before You Drink!

The way to waste less is to think more.

Suppose, for example, you want to buy some juice. In the old days, all you had to think about was which flavor to choose. But today, that is only the first step. Now you have to think about the juice container as well. What is it made of? What will happen to it when the juice is gone?

Drinks come packaged in glass, plastic, aluminum, steel ("tin") and cardboard; they come in fancy juice boxes with individually wrapped plastic straws. Which kind of packaging does the least harm to the Earth? Which is the best choice for a Trash Attacker?

These questions are surprisingly difficult to answer. When plastic juice bottles are made, the air and water are polluted. When cardboard cartons are made, the air and water are polluted, too. Which pollution is worse? It is hard to say.

Glass bottles can be recycled or reused, which often makes them a good choice. But they are also quite heavy. When they are trucked from place to place, a lot of gasoline is burned, and poisons from the exhaust are released into the air. Special juice boxes, on the other hand, cannot be recycled or reused, but they are lightweight. When they are transported, less fuel is used and less pollution goes into the air.

We don't have all the answers. We do know that buying juice in small containers makes a lot of garbage. Buying large containers is much less wasteful. This means that small juice boxes, jars, cans and cartons are OUT. Large "family-size" containers, from which you fill your own reusable glass or Thermos, are IN.

Which "family-size" container should you choose? In general, the best choice is a container that can be taken back to the store and **refilled**. If you can't get that, choose a container that can be **recycled**. (Find out which materials are recycled in your community. Choose something that's on the list.) If there's no recycling where you live, choose a container that can be **reused** at home. Do NOT buy containers that cannot be refilled, recycled or reused.

REUSE !

The second-best way to cut down on garbage is as easy as the first. Instead of throwing things out, reuse them.

All kinds of things can be saved and used again—yoghurt tubs, jam jars, plastic bags, old rags, gift wrap, buttons, nails, wire and string. Torn jeans can be mended or cut off to make a pair of shorts. Broken bicycles can be fixed. Clothes and toys that you've outgrown can be given to younger friends, or to someone else who needs them. You might even be able to sell them at a yard sale or second-hand store.

Second-hand stores, flea markets and sales are also good places to look for things you need. They're fun, they're friendly, and they're cheap. You'll be amazed by what you can buy—T-shirts, suspenders, comics, puzzles, books, bicycles, video games, tapes, skateboards, radios, skis—almost anything you can imagine! It's like a treasure hunt. Second-hand doesn't mean second-rate!

Some used things cannot be bought or sold. Nobody wants to buy old tissue boxes, for example, or empty toilet rolls. But these items can be put to good use around your school or home.

Sometimes, in order to reuse things, we have to see them in a new way. At first glance, an empty milk carton may seem like a useless piece of junk, but look again! With a little work, it could become a plant pot, a bird feeder, or a garage for toy cars. A toilet roll may be the turret of fairy castle on Monday and a runway in a marble racetrack by the end of the week. A lot of what we throw away is too interesting to waste. Things that you can't use yourself could go to a daycare center or nursery school.

The only trick to saving things for reuse is having places where they can be put away. This is simple, but it's important, too. For example, if you want to give your old magazines to the children's ward at the hospital, you need a box or bag to collect them in. Otherwise, they'll end up in a messy rumpled muddle that nobody can use. If you want your family to save juice cans for your art projects, you need to agree on a place—maybe a small box under the kitchen sink—where they can be kept. Otherwise, they'll get in everyone's way and end up in the trash. At school, your class could have a bin to collect paper with one good side, to be used for rough work. If you put the bin beside the wastepaper basket, saving paper will soon feel natural. The easier you make it, the better your system will work.

I think that I'll
Turn on my
*   brain*
And see what I
Can use again!

Refilling Is Thrilling!

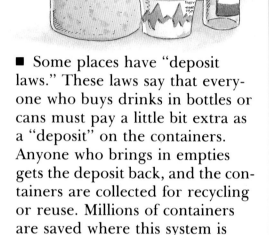

- Some stores sell milk, shampoo, soft drinks, peanut butter and other products in containers that can be refilled. Choose them whenever you can.
- In Japan, most bottles are filled and sold at least three times. Then they're recycled.
- In Denmark and Norway, only a few sizes and shapes of pop bottles are permitted by law. If all pop bottles are the same, then any bottler can refill them. If there are many different kinds, a bottler can only use those that are the right shape for the brand of pop he makes.
- Some places have "deposit laws." These laws say that everyone who buys drinks in bottles or cans must pay a little bit extra as a "deposit" on the containers. Anyone who brings in empties gets the deposit back, and the containers are collected for recycling or reuse. Millions of containers are saved where this system is used.

Trash Attack!

Reuse and Win!

Hold a contest at school to see who can make the following things entirely out of materials saved from the garbage. Shop at second-hand stores for silly prizes. Have fun!

- the best costume
- the most interesting toy
- the most enjoyable game
- the most useful item
- the most tuneful musical instrument

The Price Is Right!

Here are ways to earn your **Trash Attack!** certificate—and save money at the same time. Give yourself one point for each suggestion you try. Come up with ideas of your own and give yourself points for them, too!

■ Take your outgrown skates to a skate exchange, and buy new ones there. Wiggle your toes and enjoy the comfort of your new (used) pair. Someone else has already broken them in! Figure out how much you saved by buying second-hand.

■ When you get tired of your toys, try trading with a friend. That way you'll both get something new—for free! Or hold a toy or book exchange at school.

■ Go shopping at a garage sale or flea market. You will be amazed to see how far your money goes.
■ Set up a "scrap paper" box for your family. Ask everyone to save paper with one blank side—bulletins from school, advertising, letters, envelopes, and the like. Then use it for shopping lists, telephone messages and rough homework.

■ Learn how to care for your bicycle by yourself—to oil and adjust it, and all the rest. Find out how to fix it when it breaks. (Think how good it will feel to do it for yourself!) Take a bike-repair course, or ask your parents to show you how. The better care you take of your bike, the longer it will last. If you outgrow it, it will be in good shape for the person who buys it from you. You'll get more money for it, too, than you would if it were a wreck.

Trash Attack!
Bottle Battles

Should all drinks be sold in refillable bottles? Should deposits be charged on the containers, so that more will be returned?

Many drink manufacturers and storekeepers say "no." Refillable bottles can be a nuisance for drink manufacturers, because they have to gather them up and clean them before they are reused. Deposit laws can be nuisance for storekeepers, who have to collect the extra money and keep track of it.

But others—people who are concerned about garbage—say "yes" to both refillables and deposit laws. In some places, these people are winning; in others they are not. What is happening where you live? Call your local environmental group and find out. Ask what you can do to help.

Pop cans are a source of riches.
Do not pitch them in the ditches!
Do not throw them in the trash.
Turn them into cold hard cash.

RECYCLE!

If you can't leave the garbage at the store and can't use it anymore, then recycle.

Recycling is a kind of manufacturing that turns garbage into things we can use. Most of what we throw away is still usable; it's just in the wrong place. The wood fibers in old newspapers aren't ruined because they've soaked up a little printers' ink. Clean them, take them to a paper mill, and they can emerge as cardboard, animal bedding or insulation.

When newspapers are tied in bundles, one tonne of them takes up about the same amount of space as a double bed. Every time we recycle this amount of newsprint, it means that about 20 trees do not have to be cut down to make new paper.

We also save enough energy to run a home for more than 6 months and enough water to fill 200 bathtubs! What's more, we reduce the amount of pollution we put in the air.

Aluminum—the light, flexible metal used in beverage cans—is an even better deal. Recycling one aluminum container saves the energy of half a can of gas, which would be needed to make a new container from scratch. This is because aluminum comes from rocks, and enormous amounts of energy are needed to smelt, or separate, it. The aluminum in a drink can has already been set free, so all this energy can be saved through recycling. This explains why scrap aluminum always sells for such a good price.

For recycling to work, we have to get valuable materials from where they are—in millions of garbage cans—to the places where they can be processed, such as paper mills, aluminum smelters, composting plants and glass factories. Because each material has to be handled separately, recycling always starts with sorting: paper in this pile, glass in that bag, tin cans over there, and so on. That's where you come in. All you need is a container for each material that you're asked to separate. Instead of one garbage can, you may end up with three or four.

The next step is to get everybody's sorted materials together. In some towns, families deliver their recyclables to neighborhood depots or bins. In others, they put things out in special containers, to be picked up with their trash.

Once the used materials are neatly sorted and collected, they can be sold to factories and mills for reprocessing. This is the trickiest part of recycling. The companies that purchase used materials are

businesses. They buy used materials and sell recycled goods. The more recycled products they can sell, the more used materials they are able to buy. They cannot afford to buy more than they can use.

If many communities have collected the same materials, the factories may not need it all. This sometimes happens with newspapers and glass, for example. Prices fall and communities receive less money than they expected for their recyclables. The amount they receive may not be enough to pay for collecting the materials, storing them and transporting them to the recycling plant. People may start to get discouraged. Some of them throw up their hands and decide that recycling doesn't work.

Don't believe it! Recycling is a new idea in an old, throw-away world. Those old ways will have to change. For example, think what a difference it would make if everybody decided to Buy Recycled. We could have labels in the stores to show which items contained recycled materials. And what if our governments—all the governments in the world—announced that they were going to Buy Recycled. Together, we could change everything. The mills and factories would be able to buy all our used materials and work full blast at recycling.

Governments can help in other ways, too. Right now, businesses that produce new materials, like mines, oil wells and big lumbering companies, often get money and tax breaks from the government. Companies that sort, collect and process used materials generally don't get the same help. But they could. Governments just need to change their rules.

Governments work for us and spend our money. If enough of us ask for changes, governments have no choice but to pay attention.

Why couldn't we have lower taxes for businesses that buy recycled goods? Or a "garbage sales tax" that is added to the price of everything we buy? This money could then be paid back to people who sort their garbage for recycling. There are a lot of things we can try.

Recycling is a smart idea. If we are smart, we will make it work!

Sort-It-Yourself

There are two methods of sorting recyclables. One is called "source separation." This is what we are doing when we sort our garbage at home. We are the "source" of used materials and we are "separating" them—glass here, paper over there —so they can be sold and recycled.

But in some communities, people do not sort their garbage themselves. Instead, all their recyclables are mixed together. This garbage is then collected and taken to a central plant where it is sorted by machines and by paid staff. This method is more expensive and less effective than source separation: more recyclables are missed. There is also more "contamination"—beans stuck inside cans, staples left in paper, newspapers soaked with

Join in the Recycling chorus! Save some paper! Save the forests!

cooking oil, and other problems of that sort. This lowers the quality of the materials and makes them harder to sell and to process.

The only way to make sure your recycled goods are not contaminated is to check them carefully yourself.

Exciting Recycling!

■ Recycling always uses less energy and causes less pollution than making goods from new materials.

■ Recycling usually costs less than other ways of getting rid of trash. Recycling aluminum cans is always a money-maker.

■ Recycling saves forests, petroleum and other resources.

The Garbage Glossary:

RECYCLING WORDS

Contamination (kon-ta-mi-NAY-shun): Materials mixed in with recyclables that lower their quality. Examples are food scraps left in cans, tape and plastic stuck to paper and coated paper put in with newsprint.

Non-recyclable (non-ree-SIKE-la-bull): Materials that are difficult or impossible to recycle. This includes some plastics and all products that contain more than one material. Envelopes with plastic windows, aerosol cans and lightbulbs are all non-recyclables.

Recyclable: Materials that can be recycled. This includes paper, glass, steel, aluminum, food scraps, some plastics and used motor oil.

Source separation: Sorting recyclables at home.

Problems With Plastics

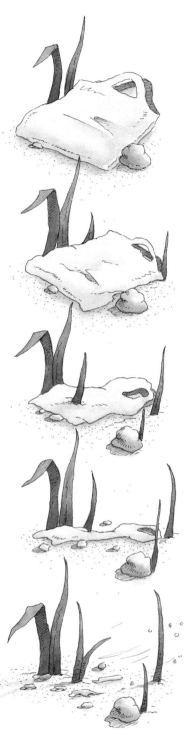

Plastics are difficult to recycle. For one thing, they are bulky—a little bit of material takes up a lot of space. This makes them expensive to transport.

For another, they are fussy to work with. Not all plastics are the same. PET, the material in a plastic soft drink bottle, is very different from PVC, which is the plastic used to make black plastic pipe. There are hundreds of other kinds as well, each with its own special use. How could we sort out all the different kinds for recycling? To make matters even more confusing, one item (a squeezable plastic shampoo bottle, say) may contain layers of several different plastics—a certain one for strength, another to keep out air, another for flexibility, and so on.

When different kinds of plastics are recycled together, they lose their special qualities. Still, they can be made into useful things such as plastic lumber, pilings for wharves, fenceposts that will never rot, unless ...

This "unless" is a new problem. Someone has just invented a kind of plastic that does rot. Well, that *sort of* rots. Plastic is combined with starch (the kind you have in your kitchen) to make "biodegradable plastic." When the starch rots, the plastic crumbles into tiny bits like dust. Another kind, called "photodegradable plastic," breaks down in sunlight. In both cases, the flecks of leftover plastic are a new and unknown form of pollution. So it is safest to avoid both of these new plastics.

Not only that, but neither biodegradable nor photodegradable plastic can be recycled. And if they are accidentally added to recyclable plastic, the end product is ruined.

Too Good to Be True?

Here are some statements about recycling. Which do you think are true? Which sound like tall tales?

■ Apartment buildings are being built with separate garbage chutes for collecting different kinds of recyclables.

■ Instead of old-fashioned garbage trucks, towns can buy recycling trucks, with compartments for various materials.

■ At some landfills, dumping is free if you have sorted your garbage for recycling. Otherwise, you have to pay a "tipping fee," or you may not be allowed to dump your garbage at all.

■ In some places, it is against the law to throw recyclables in the trash. The garbage collectors check, and anyone who cheats gets a ticket.

■ Some garbage cans have two compartments—one for food scraps and one for dry trash. Two-compartment garbage trucks are made to go with these cans. Special lifters on the back bumper pick up the cans and dump the garbage so that wet and dry fall on different sides of the truck.

Food scraps collected in this way are made into compost.

■ In some places, people pay a bit extra when they buy soft drinks, as a "deposit" on the cans. The money is refunded when they bring the cans back. Almost ALL the cans are returned for recycling.

■ You know about machines that sell pop. There are now machines that take back empty cans and give people their deposits.

■ Cities and towns in Pennsylvania, U.S., have been told to sort all the trash in all their landfills to find recyclables.

■ Billions of old tires lie in North American dumps. Up until now, very few of them could be reused. But now a way has been found that could turn all old tires into plastic pipe, fabric and new tires.

■ In some places, people who bring in their newspapers for recycling are given a week's supply of paper products, such as tissues and toilet paper, in return.

Good news! All these statements are TRUE!

Compost Is Not a Dirty Word

Much of the food we eat—tea, coffee, fruits and vegetables—comes from plants. When plants rot, they turn into food for the next season's crops. But in the garbage, all this natural fertilizer goes to waste.

The solution is simple: compost. Composting takes kitchen scraps, grass clippings and leaves, and turns them into rich, crumbly garden soil. Some cities compost huge amounts of household waste and use it to fertilize their boulevards and parks. Composting is also done by businesses that produce high-grade topsoil for sale.

Do-it-yourself composting is fun. A well-con-

structed compost heap is made of layers: first manure or compost starter (which you buy at a garden store), then kitchen scraps, then dirt, more scraps, more dirt and so on. You can put in fruit and vegetable peels, egg shells, tea and coffee grounds and grass clippings; chop them into small pieces for fast results. Do not add meat, bones, dairy products or foods with butter or oil on them, because they will make your compost smelly and attract animals.

Keep your compost moist—like a squeezed-out sponge—and turn it with a shovel once or twice a week. You'll have finished compost in six months to a year. Dig it into your garden, and watch things grow! There is no better fertilizer than compost.

The Lazy Gardener's Notebook

If you don't have manure or compost starter, don't worry. It's not even necessary to layer your heap. You may not even need to turn and water it. Here's an easy method that often works, so why not give it a try? Collect your kitchen scraps in a heap for several months. In the fall, spread them on the garden, dig them in and leave them there. In the spring, plant as usual. You might dig up an egg shell once in a while, but who cares?

Refuse to Make Refuse!

Here are some things that you can do to make recycling work. For each suggestion you follow, give yourself one credit toward your **Trash Attack!** certificate.

you live, maybe it's time there was. Talk to your parents, friends, teachers and group leaders. Find other people to work with you; you will need at least one adult to help. In some towns, groups of children—choirs, clubs, schools, teams—collect old newspapers or aluminum tins to raise money. Don't worry if you can't recycle everything right away. Even a small beginning is a big step in the right direction.

■ When you prepare materials for recycling, follow the rules carefully. Rinse containers. Remove staples and paper clips from paper. Tie newspapers with twine, not plastic cord. Careful recyclers make for good-quality recycled goods.

Recycle

■ If recycling has begun in your community, find out how it works. Your friends and neighbors may know. If not, call city hall. NOW is always the right time to start recycling.

■ If there is no recycling where

Buy Recycled

■ Look for recycled paper and other products in the store. Buy them whenever you can.

■ Choose eggs and other food in paper packages instead of plastic. Egg cartons and cardboard trays are often made from recycled paper.

■ Try not to buy products that cannot be recycled.

Speak Out

- If you cannot find recycled products in your stores, talk or write to the manager.
- If your town has not started recycling, or if some recyclables are not included, write to the mayor and your local newspaper.
- If your government has not announced that it will Buy Recycled, write to the head of the government and complain. Better yet, see if your class or club wants to write letters as a group project.
- If there is an environmental society or young naturalists' club in your community, join it. Garbage is too big a problem to take on by yourself.

Thinking about a big problem, like garbage, sometimes makes us feel very small. "People are throwing things out everywhere, all the time," you may think. "How can one little person change it all?" It is not just children who feel this way. Adults often do as well.

But the fact is that what you do is important. Many people, working together, can make things change.

You are not alone. Millions of people, all around the world, are working to keep the Earth green and growing. They are buying less, reusing what they already have, and recycling—just like you. You cannot be the whole answer, but you are part of it!

Treasures In Your Trash

(All the materials in this chart can be recycled.)

Material	One person in one year throws out …	Found in …	To recycle this material …	Recycled materials are used for …	The good news is …
Food scraps and yard wastes	30 cans	Fruits and vegetable peels, grass clippings, leaves.	The material is mixed with soil, turned now and then to provide it with air, and watered from time to time. It turns into rich soil called "compost."	Garden soil and fertilizer.	Nutrients that came from the soil are put back again.
All paper	25 cans	Newspapers, boxes, writing paper, computer paper and so on.	The ink is removed through a special washing process. The paper is mashed into pulp, mixed with new pulp made from logs, and used to make paper.	Newsprint, boxes, insulation, wall-board, cat litter, egg cartons, cardboard trays, tissue, writing paper.	Printing the Saturday edition of a big-city newspaper on recycled paper saves more than 10,000 trees.
Glass	5 cans	Jars and bottles.	The glass is crushed to make "cullet." It is mixed with sand, limestone and soda ash and melted in a hot furnace to make new glass.	New containers, fiberglass, pavement.	If you make a bottle from recycled glass instead of from scratch, you will save enough energy to light a lamp for 4 hours.

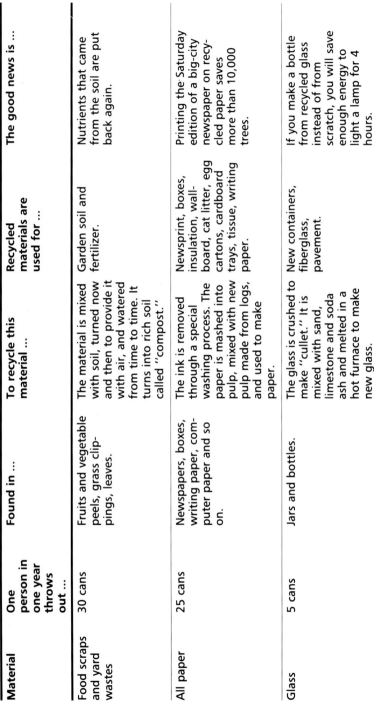

Steel	5 cans	"Tin" cans—the kind that food such as tinned soup comes in. (They are actually made of steel, with a coating of tin.)	Cans are melted with iron ore in a blast furnace when new steel is being made. The tin coating is saved, too, and used in making car parts.	New cans, cars, refrigerators, siding, tools, stainless steel and many other things.	Recycling one ton of steel saves $1\frac{1}{2}$ tons of iron ore and 3 to 4 barrels of oil.
Aluminum	1 can	Beverage cans	Cans are shredded and decorations are removed. The cans are then melted to make a sheet of pure aluminum.	New aluminum cans.	Recycling one can saves the energy equivalent of half a can of gas.
"PET" plastic	1 can	Large pop bottles	The bottles are shredded. Some are made into pellets. Others are re-manufactured to make new plastics.	Stuffing material, filters, handles, paintbrush bristles, plastic lumber, insulation.	Recycling plastic saves petroleum, a non-renewable resource.
Used motor oil		Leftover from oil changes	The oil is taken to an oil refinery and refined again.	New motor oil.	Used motor oil is a Hazardous Waste. Recycling saves the leftover oil and keeps it out of incinerators and landfills.

TOTAL RECYCLABLES = 67 cans, or about 80 percent of your garbage!

Index